**Juliana de Carvalho Passos**
**Davi Lima**
**Daniel Liarte**

# Identification of phenotypic and molecular markers for obesity

**Juliana de Carvalho Passos**
**Davi Lima**
**Daniel Liarte**

# Identification of phenotypic and molecular markers for obesity

**Final course work presented to the Nutrition course at the Federal University of Piauí**

**Imprint**
Any brand names and product names mentioned in this book are subject to trademark, brand or patent protection and are trademarks or registered trademarks of their respective holders. The use of brand names, product names, common names, trade names, product descriptions etc. even without a particular marking in this work is in no way to be construed to mean that such names may be regarded as unrestricted in respect of trademark and brand protection legislation and could thus be used by anyone.

Cover image: www.ingimage.com

This book is a translation from the original published under ISBN 978-613-9-69162-3.

Publisher:
Sciencia Scripts
is a trademark of
Dodo Books Indian Ocean Ltd. and OmniScriptum S.R.L publishing group

120 High Road, East Finchley, London, N2 9ED, United Kingdom
Str. Armeneasca 28/1, office 1, Chisinau MD-2012, Republic of Moldova, Europe
Printed at: see last page
**ISBN: 978-620-8-12607-0**

# INDEX

# CHAPTER 1

## INTRODUCTION

Obesity is a disease characterised by the excessive accumulation of body fat at a level that compromises the health of individuals, causing damage, mainly related to metabolic alterations, breathing difficulties and locomotion (WANDERLEY and FERREIRA, 2010).According to the World Health Organisation, a parameter used to diagnose obesity is the *body mass index* (BMI), obtained from the ratio between body weight (kg) and height $(m)^2$ of individuals. Individuals whose BMI is equal to or greater than 30 kg/m$^2$ (WHO, 2000) are considered obese.

It is known that obesity can be caused by a combination of factors, including environmental and genetic aetiology. Although genetic obesity requires multi-professional monitoring that is different from environmental aetiology, there are insufficient efforts to personalise treatment. The number of genes associated with obesity is increasing and their diagnosis requires expensive and complex tests. For obese people, however, knowing the causes of their condition is fundamental for a better prognosis (RODRIGUES, 2016) (FRIGERI, 2015).

Due to the lack of a precise diagnosis, obese patients are most often subjected to diets which, even if they change their eating habits and practise physical activity, do not produce satisfactory results and can worsen their obesity, due to the "compensation" effect of the food restriction to which they have been subjected. In more serious cases, some patients are referred for bariatric surgery and after some time return to the weight they had before the procedure (PROVENCHER, 2003) (DAYYEH, LAUTZ and THOMPSON, 2011) (BASTOS, et al, 2013).

The investigation of genetic obesity is usually based on a family history interview. However, due to limitations, this interview does not allow a diagnosis to be formulated and genetic testing is necessary. This genetic test is done through DNA sequencing, which is a very expensive and complex technique that is not often used in clinical routine (FRIGERI, 2015) (CONTI, MORENO and ONG, 2010). That's why there's a need for new tests that are cheaper and that support the nutritionist's behaviour.

From this perspective, the association of phenotypic and hereditary markers and the use of bioinformatics and molecular tools directs nutritional management towards monitoring the genetic individualism of each patient, allowing for the creation of more specific and individualised diet plans.

# CHAPTER 2

## THEORETICAL FRAMEWORK

### 2.1 General aspects of obesity

According to the WHO, obesity is an abnormal or excessive accumulation of body fat that can reach levels capable of affecting health (WHO, 2000). According to the Ministry of Health, it can be understood as a multifactorial condition involving biological, historical, ecological, economic, social, cultural and political issues (BRASIL, 2006a).

Despite the seriousness of this problem, there has been a significant increase in the number of people suffering from this morbidity. In the global context, on a decreasing scale, South Africa has the highest prevalence of overweight (65.4 per cent), followed by Russia (59.8 per cent), Brazil (52.4 per cent), China (25.4 per cent) and India (11 per cent) (BRASIL, 2015).

In Brazil, according to a report published by the Food and Agriculture Organisation of the United Nations (FAO) and the Pan American Health Organisation (PAHO) (2017), more than half of the Brazilian population is overweight and obesity already affects 20% of adults. According to this document, overweight in adults rose from 51.1 per cent in 2010 to 54.1 per cent in 2014. In 2010, 17.8 per cent of the population was obese; in 2014, the rate reached 20 per cent, with the highest prevalence among women, 22.7 per cent. Corroborating this data, the Brazilian Association of Obesity and Metabolic Syndrome - ABESO (2010) presented an epidemiological approach subdivided by region of Brazil, in which it was observed that the southern region (56.08%) had the highest rate of obese adults, followed by the southeast (50.45%), midwest (48.3%), north (47.2%) and finally the northeast (44.45%). São Luís and Florianópolis are the Brazilian capitals with the lowest rates of overweight (46%) and obesity (14%), respectively (BRASIL, 2017b).

According to BRASIL (2017a), the rate of overweight and obesity in the Brazilian population has increased by 23 per cent in the last nine years, and currently 52.5 per cent of Brazilians are affected by this morbidity. With regard to gender, there is a higher prevalence among males (56.5%). Among young people, there was a lower prevalence (31.5%), while the population aged 35 to 64 had higher rates (61.8%). It was also observed that the lower the level of schooling, the higher the BMI of the population; in individuals with

0 to 8 years of schooling, the prevalence of excess weight corresponded to 58.9%. In contrast to the results of BRASIL (2017), the IBGE (2015) in a survey showed that the rate of obese Brazilians is close to 60% and it was observed that the prevalence of overweight was higher in females (58.2%) than in males (55.6%).

This statistically demonstrated high prevalence of obesity is strongly linked to the nutritional transition that Brazil has been experiencing for around 50 years. According to Popkin et al, 1993, "nutritional transition is a process of sequential changes in the pattern of nutrition and consumption that accompany economic, social and demographic changes and changes in the health profile of populations". This transition is mainly characterised by changes in dietary patterns, an increase in sedentary lifestyles and the inclusion of women in the labour market. These associated factors have led the country to reduce the prevalence of malnutrition and sharply increase obesity rates regardless of age, gender or social class, which as a consequence has led to a significant increase in individuals affected by chronic non-communicable diseases (SOUZA, 2017).

Obesity is commonly diagnosed and characterised based on anthropometric parameters, which are: BMI greater than or equal to 30kg/m$^2$ , waist circumference (WC) greater than or equal to 94 cm for men and greater than or equal to 80 for women, which is more suitable for diagnosing visceral adiposity and is less related to cardiovascular complications, and waist-to-hip ratio (WHR), which, because it includes muscle tissue, is more strongly associated with insulin resistance. It is recommended to use at least these two parameters in combination to reliably characterise an obese individual (ABESO, 2009) (MARTINS and MARINHO, 2013). Obesity is a chronic non-communicable disease that is currently being considered a major public health challenge, as it causes various health risks and is closely related to metabolic syndrome (CANTALICE, et al, 2015), which according to the Brazilian Society of Endocrinology and Metabology (2016) can be described as:

> A set of metabolic risk factors that manifest themselves in an individual and increase the chances of developing heart disease, stroke and diabetes. They are based on resistance to the action of insulin. Risk factors include the presence of three or more of the factors listed below: WC greater than 102cm for men and 88cm for women men waist greater than 102cm and in women, HDL less than 40mg/dl in men and 50mg/dl in women, triglyceride equal to or greater than 150mg/dl, blood pressure equal to or greater than 135/85 mmHg and glucose equal to or greater than 110mg/dl.

One of the most common and unfavourable outcomes of obesity linked to metabolic syndrome is the development of cardiovascular diseases (RICARTE, 2017). As well as leading to high mortality rates, these diseases can cause various partial or total physical and/or mental limitations (DE MORAIS, 2017).

In view of the above, it can be seen that obesity is a serious health problem that requires effective treatment. However, it is necessary for the professional nutritionist to present a diagnosis capable of distinguishing the etiological factors of obesity and thus offer a more targeted and specific treatment to the patient.

## 2.2  Dietary obesity and genetic obesity

The etiology of obesity is generally multifactorial. It includes dietary and genetic factors. These have completely different definitions. Dietary obesity stems from an imbalance between calorie intake and expenditure and is strongly associated with poor eating habits, greater consumption of high-fat, high-calorie foods such as ultra-processed carbohydrates and *fast food;* socio-economic factors; *a* sedentary lifestyle and psychological disorders (ROMAGNA; SILVA; BALLARDIN, 2010), as well as involving factors such as excess energy and especially lipids, thus favouring an increase in adiposity and causing biochemical changes. In this context, it can be seen that nutritional monitoring and physical exercise are tools capable of reversing the clinical picture of obesity, as they can positively influence body composition by promoting an increase in total energy expenditure, a balance in the oxidation of macronutrients and the preservation of lean mass. The effects on energy metabolism will depend on the type, intensity, duration and frequency of the exercise, as well as the maintenance of the diet plan (MONTEIRO, RIETHER and BURINI, 2004). In more serious cases, in which these patients undergo bariatric surgery, the chance of surgical success is much greater, due to the very etiological characteristics of dietary obesity which, when coupled with nutritional monitoring and physical exercise, promote satisfactory results, for the same reasons described above. It is worth pointing out that weight loss is considered one of the main parameters for defining the success of bariatric surgery, since after losing weight there is a proven improvement in comorbidities (GLOY, et al, 2013) (NOVAIS, et al, 2010).

Genetic obesity investigates possible dysregulations in the hunger and satiety mechanism, hormonal alterations and specific metabolic alterations that may be associated with genetic obesity (DAMIANI; DAMIANI; OLIVEIRA, 2002). It is subdivided into monogenic and polygenic obesity. The former is defined as obesity resulting from the mutation or deficiency of a single gene. Polygenic obesity, on the other hand, is the most commonly manifested genetic influence, giving certain individuals a susceptibility resulting from genetic factors that can interrelate in a very complex way (COUTINHO and DUALIB, 2007).

Neuroendocrine factors have been identified as the most important for maintaining the body's energy balance. With regard to the neuroendocrine control of energy metabolism, research has identified the peptides leptin and insulin as the two major adiposity signalling agents that inform the brain of the amount of body energy stored as excessive (LANDEIRO and DE CASTRO QUARANTINI, 2011). These peptides are hormones secreted in proportion to fat mass and act peripherally, stimulating catabolism, but many obese individuals are resistant to these hormones. In the central nervous system, insulin and leptin interact with hypothalamic receptors, favouring satiety (HALPERN, RODRIGUES and COSTA, 2004). The process of action to be described is considered to be one of the most obvious functions that leptin has in the human organism.

> The action of leptin in the hypothalamus occurs specifically in the arcuate nucleus, stimulating and/or inhibiting orexigenic and anorexigenic neurons, thus controlling hunger and satiety mechanisms, as well as food intake and energy expenditure mechanisms. Low concentrations of plasma leptin lead to increased food intake, hunger and decreased energy expenditure, while high concentrations cause decreased food intake, satiety and increased energy expenditure (RIBEIRO, et al, 2007).

Leptin acts when bound to its LEPR receptor, which has a transmembrane domain similar to that of the cytokine receptor family. However, genetic variations affecting the protein or signalling sites can influence this mechanism (FRIGERI, 2015). In a study by Rojano Rodriguez (2016), results showed that the rs1805134 polymorphism could be involved in the development of morbid obesity and it was also observed that several LEPR mutations have been described in patients with early onset of severe obesity and hyperphagic eating behaviour. From this perspective, it can be concluded that genetic obesity cannot be controlled by dietary re-education combined with physical activity alone. In these patients, bariatric surgery may not be an efficient alternative for weight loss, since the mechanism involved in the process includes "internal" factors inherent to the individual's metabolism and genetic material.

Identifying the etiology of obesity would enable nutritionists to offer obese individuals adequate nutritional support. However, this aetiological distinction is little explored in the professional routine.

## 2.3 Genes associated with obesity

Since the conclusion of the human genome project, a growing number of polymorphisms associated with genetic obesity have been discovered, which has enabled new tools for analysis and study. Fernandes, Fujiwara and Melo (2011) presented a list of 8 genes that have polymorphisms associated with genetic obesity. While Frigeri (2015) suggests in his study that 9 genes may have obesity-related polymorphisms, a technical report developed by the company GenoVive (2011) listed 20 genes with the same association mentioned above. Another study by Ochioni (2016) investigated the relationship between 12 polymorphisms of inflammatory mediator genes involved in obesity. Among the most cited genes in the literature are FTO (Fat mass and obesity associated), MC4R (Melanocortin 4 receptor), LIPC (hepatic lipase) and TCF7L2 (Transcription factor 7-like 2).

The FTO gene encodes a protein that affects the hypothalamic region of the brain that regulates appetite, energy intake and satiety and is associated with monogenic obesity and hyperphagia, thus being involved in appetite regulation. The most widely studied FTO polymorphism (T>A) shows that the A allele is the risk allele for being overweight (FERNANDES; FUJIWARA; MELO, 2011). While the polymorphism (A>G) of the MC4R gene seems to predispose post-menopausal women to developing some symptoms of metabolic syndrome.

It is known that this gene may also be associated with monogenic obesity and appetite regulation (BRODOWSKI et al, 2017). The LIPC gene catalyses the hydrolysis of triglycerides and phospholipids in plasma lipoproteins, contributing to the remodelling of low-density lipoprotein remnants VLDL, LDL and HDL (VERMA et al, 2016). The TCF7L2 gene is a transcription factor present in glucose metabolism that is significantly associated with susceptibility to type 2 diabetes mellitus (WANG et al, 2013).

In general, polymorphisms and mutations associated with obesity are identified through DNA sequencing, as this is the best-known way of identifying polymorphisms present in a DNA fragment amplified by PCR (Polymerase Chain Reaction). In his work, Frigeri (2015) describes in detail how sequencing is carried out. This aforementioned method indicates how the repeats are present in the microsatellite markers and which allele is in the SNP (Single Nucleotide Polymorphism) (PLOMIN, et al, 2016).

However, this sequencing technique is not carried out in every laboratory due to its complexity and high cost, making it inaccessible to the majority of the population. It is also little known among nutrition professionals.

## 2.4 Nutritional monitoring of obese people and the prospects for genetic studies

Dietary treatment for obese patients aims to achieve weight loss through lifestyle changes, focusing mainly on the individual's dietary re-education. Dietary treatment for these patients is usually based on hypocaloric and restrictive diets, because according to Lopes (2017) hypocaloric diets promote caloric deficits in intake, inducing greater use of lipids as a source of energy substrate, thus promoting positive changes in the body composition and metabolism of these individuals (LOPES, 2017) (MACHADO and KIRSTEN, 2016).

The effects of low-calorie diets can be seen in some studies, for example, Rocha, Moreira and Boroni (2016) showed in a study that nutritional assessments carried out after prescribing low-calorie diets showed a reduction in measurements such as weight, body mass index, waist circumference, waist-to-hip ratio and body fat. Another study by Lopes (2017) corroborates this data, in which the main findings were that after the intervention period obese individuals showed a reduction in body mass, a reduction in total and visceral body fat, with no changes in fat-free mass.

Although the primary goal of dietary treatment for obesity is to reduce body size and weight, it is important to plan a diet that is appropriate and balanced, that includes all the food groups and that respects the patient's individual characteristics (BRASIL, 2006b) (ROCHA and MOREIRA, 2016).

However, it is known that there is a significant proportion of obese individuals who, even with this type of dietary treatment, fail to achieve their weight reduction goals and consequent metabolic improvement. It is possible that anthropometric and biochemical analyses are not enough to take into account the individual and metabolic characteristics of these individuals. From this perspective, biases arise to propose a more specific and personalised analysis of this patient, taking into account their genetic material. Only with the emergence of post-genomic knowledge in 2003 was it possible to observe a rise in the field of nutritional practice, as knowledge emerged with enormous potential to change the future of dietary guidelines and personal recommendations. In this context, sciences have emerged that are revolutionising the future of nutrition: nutrigenetics and nutrigenomics. The former refers to the interactions between dietary habits and the genetic profile of each individual (SCHUCH; VOIGT; MALUF; ANDRADE, 2010) (ORDOVAS; CORELLA, 2004).

It is hypothesised that these different responses are associated with the presence or absence of specific biological markers, usually genetic polymorphisms, which could then predict the individual's response to the diet. This would make it possible, in the future, to prescribe a "personalised diet" according to the patient's genetic makeup (ORDOVAS; CORELLA, 2004).

Nutrigenomics is a science that studies how food constituents interact with genes and their products in altering phenotypes. To this end, it is necessary to understand how nutrients and bioactive compounds act in modulating gene expression (FUJII; MEDEIROS; YAMADA, 2010).

In addition to nutrition, other sciences are fundamental to the development of nutrigenomics and nutrigenetics, such as bioinformatics, which is the union of software engineering, maths, statistics, computing and molecular biology (PROSDOCIMI, et al, 2002).

# CHAPTER 3

**OBJECTIVES**

## 3.1 GENERAL OBJECTIVE

To identify phenotypic and molecular markers for obesity and assess the feasibility of using these markers in the multiprofessional monitoring of obese patients.

## 3.2 SPECIFIC OBJECTIVES

- Select phenotypes associated with obesity and check for indicators of genetic inheritance;
- To identify genetic polymorphisms associated with obesity and its genomic characteristics, with a view to using them in molecular diagnostic tools;
- Develop molecular analysis tools for the polymorphisms identified, looking at their technological and economic viability;

## CHAPTER 4

**METHODOLOGY**

To achieve the objective proposed in this work, three different approaches were used: identification of phenotypic markers (based on symptoms and comorbidities associated with genes involved in genetic obesity); analysis of genetic patterns suggestive of heredity (construction and analysis of heredograms); computational and molecular analyses involving the prediction of molecular diagnostic tools and experimental training. The phenotypic and hereditary approaches, in association, are intended to help nutritionists classify obese patients into a risk group for genetic obesity or not, leading to a more in-depth aetiological investigation. Molecular analysis helps choose the genetic test to be applied in each case.

### 4.1 Phenotypic and hereditary analysis for genetic obesity

At this stage, a survey was carried out on the genes associated with genetic obesity and the symptoms, conditions or diseases (collectively known as phenotypes) that these could cause. Based on this, a form was drawn up which sought to analyse the phenotypic expression of the individuals in order to identify whether the characteristics displayed could be related to genetic factors. In addition, the form also analysed the hereditary profile through questions investigating the degree of obesity in the ancestors and direct descendants of these individuals. The form was distributed via social networks over 6 days to a specific group of obese individuals who had undergone bariatric surgery. In order to answer the questionnaire, they had to fill in the Informed Consent Form, which was provided before the questions.

The answers were used to construct heredograms, which were analysed according to Mendelian genetics. After analysing the hereditary profile among individuals who had undergone bariatric surgery, this finding was compared with post-surgical success. With regard to phenotypic characteristics, the sample was divided into two groups, considering the individuals who reported weight gain after the post-surgical process and in another group those who did not have weight gain, and the phenotypic signs and symptoms that occurred most in each of these groups were identified.

## 4.2 Predictive computer analysis

## 4.2.1 Data mining: Identification of polymorphisms associated with obesity.

This stage used a computational approach known as *data mining,* which according to Fayyad, Shapiro and Smyth (1996) is the non-trivial process of identifying valid, new, potentially useful and ultimately understandable patterns in data. To this end, a bibliographical and molecular data survey was carried out on the main polymorphisms associated with genetic obesity. From this survey, the biological sequences corresponding to the polymorphic regions of interest were downloaded and organised in a local database for subsequent analysis. The main databases used are listed in Table 1:

**Table 1:** Main bibliographic and molecular biology databases used for genetic obesity data mining.

| Name/Email address | Description |
|---|---|
| Pubmed* https://www.ncbi.nlm.nih.gov/pubmed/ | Search engine with free access to the MEDLINE database of citations and abstracts of research articles in biomedicine. |
| Scielo- Scientific Electronic Library http://www.scielo.org | It is an electronic library covering a selected collection of Brazilian scientific journals. |
| Capes Journal Portal http://www.periodicos.capes.gov.br | Virtual library that brings together and makes available to teaching and research institutions in Brazil the best of international scientific production |
| DbSNp* https://www.ncbi.nlm.nih.gov/snp | Database of single nucleotide polymorphisms (SNPs) and multiple small-scale variations including insertions/deletions, microsatellites and non-polymorphic variants. |
| ClinVar* https://www.ncbi.nlm.nih.gov/clinvar/ | This bank aggregates information on genomic variation and its relationship with human health |
| OMIM* https://www.ncbi.nlm.nih.gov/omim | It is a comprehensive and authentic compendium of human genes and genetic phenotypes that is freely available and updated daily. |
| Gene* | It includes nomenclature, reference sequences (RefSeqs), maps, pathways, variations, phenotypes and links to genomic, phenotypic and locus-specific resources |

| https://www.ncbi.nlm.nih.gov/gene | around the world. |
|---|---|
| Nr and Human Genome database for BLAST[+] https://blast.ncbi.nlm.nih.gov/Blast | Database containing information pertinent to the human genome that was used to compare with sequences of polymorphic genes for obesity |
| REBASE, The Restriction Enzyme. Database http://rebase.neb.com | Comprehensive database of information on restriction enzymes, |

* Databases available on the NCBI portal (http://www.ncbi.nlm.nih.org).

[+] Database used only via online access.

### 4.1.1 Structural annotation: identification of conserved and variable sites

Once the main polymorphisms associated with genetic obesity had been identified, the genetic characteristics of the DNA sequences involved were investigated, in a process known as structural annotation. To check for the existence of other polymorphisms associated or not with obesity and close to the sequences of interest, fragments of 1001 base pairs on average (500 nucleotides 5' and 500 nucleotides 3' of the polymorphism) were compared with a database of the human genome, using the BLAST (Basic Local Alignment Search Tool) programme (ALTSCHUL et al., 1997). As well as checking for other polymorphisms, BLAST made it possible to investigate whether there were other genetic sequences with similar characteristics to the target sequence in other regions of the genome (such as duplicated genes and repetitive DNA). After BLAST analysis, all the sequences with similar characteristics to the region under study were downloaded and aligned using the Clustal 2.1 programme (THOMPSON, 1997). This program "forces" a positioning of the nucleotides of the analysed sequences according to their similarity, making it easier to identify conserved and variable regions, as well as different types of mutations (substitutions, InDels, etc.).

### 4.1.2 Constructing restriction maps and selecting genetic markers

After the structural annotation, a predictive analysis of the restriction profiles of the polymorphic sites associated with genetic obesity was carried out. To do this, restriction maps were constructed using the REBASE database (ROBERTS, et al, 2015) and the Bioedit programme (HALL, 1999). Among the criteria adopted for selecting the enzymes were: enzymes that cut the sequence up to five times, regardless of the size of the enzyme's recognition site or whether it was degenerate. In addition, enzymes from all the manufacturing companies and all the schizomers available in the database were selected for analysis. The restriction maps of the sequences containing the different polymorphisms identified were compared and the enzymes capable of generating distinct profiles associated with these polymorphisms were selected.

### 4.1.3  Design of primers and virtual gel

Having identified the genetic polymorphisms associated with obesity and the restriction profiles that indicate these polymorphisms, the next step was to design the oligonucleotide primers for the Polymerase Chain Reaction (PCR). PCR is a well-established molecular biology technique that is easy to carry out and relatively inexpensive (compared to other molecular techniques). These factors make PCR combined with the analysis of enzyme digestion profiles (a technique known as PCR-RFLP) a promising molecular diagnostic tool.

The GeneTool programme (LYON, 2000) was used to design the primers. As well as helping to identify the best primer oligonucleotides, it also builds virtual gels based on the profile of DNA fragments generated by the product amplified from these primers and digested by specific restriction enzymes.

### 4.1.4  Selection of biomarkers based on the technical and economic feasibility of computationally predictive tools

Based on the data from this research, molecular analysis tools were selected to identify genetic obesity factors, considering lists of polymorphisms, restriction profiles and PCR primers. In addition to the data described above, other aspects were considered, such as the frequency of polymorphic alleles in the population, the economic cost and ease of acquisition of the materials needed to carry out the techniques, as well as the possibility of integrating multiple techniques into a single analysis protocol.

### 4.3 Experimental analysis

In parallel with the computational research, experimental protocols were evaluated in order to optimise the cost/benefit ratio for the techniques developed. The main techniques involved were DNA extraction, polyacrylamide gel electrophoresis, PCR and enzymatic digestion. However, it is important to emphasise that the last two protocols could only be carried out after acquiring the oligonucleotide primers and restriction enzymes that were identified during the project.

### 4.3.1 Sample collection and DNA extraction

The sample was collected using an oral swab. Volunteers were instructed to swab 100ml of distilled water beforehand and the sample was collected by scraping the inside of the cheeks with small sterile cytological brushes, making circular movements approximately 30 times; the brushes had the outer portion of the shafts cut off and placed in 2ml microtubes. The samples were stored in the fridge for 2 to 30 days before extraction.

DNA was extracted from the samples using NaCl. 200pl of TES (Tris HCl 10mM pH 7.6; EDTA 1mM; SDS 0.6%) and 5pl of proteinase K (10mg/ml) were added to the tubes containing the swab and incubated for 2h at 42oC. After incubation, the brush was removed and a volume of approximately 250pl was obtained, to which 42pl of saturated NaCl (6 M) was added, shaking vigorously by hand. It was centrifuged for 1 minute at 15,000 x g. The supernatant was transferred to a new tube and 2 times the volume of absolute ethanol was added. The tubes were shaken and centrifuged for 1 minute at 15,000 x g. The absolute ethanol was discarded and 1ml of 70% ethanol was added, inverting the tubes several times to wash the pellet. The tubes were centrifuged for 1 minute at 15,000 x g and the supernatant discarded. Washing with 70% ethanol was repeated once more, after which the supernatant was discarded and the tubes left open for 30 minutes to evaporate the residual ethanol. The DNA was dissolved in 60pl of TE 10:0.1 (Tris HCl 10mM; EDTA 0.1 mM).

### 4.3.2 Polyacrylamide gel electrophoresis

To create a good gel, the spacers, plates and comb were first cleaned with alcohol. The plates and spacers were assembled avoiding unevenness and using supports. The polyacrylamide gel was prepared at the desired concentration, the gel was applied to the support using a pipette and then the comb was placed to avoid air bubbles. The gel was then checked for polymerisation (10 minutes), the comb was removed and the channels were washed with MiliQ water to remove excess acrylamide, using a needle for the pieces of gel. The support was adapted to the vat and new TBE 1X (Tris, boric acid and EDTA) was placed on the inside and used TBE on the outside. Under a larger glass plate, film paper of a size relative to the number of samples that had been prepared was placed in the centre, 3pl of the 2x sample buffer (bromophenol blue and xylene cyanol) was poured in and then 3pl of the sample was poured in and homogenised, the sample and the 0x standard were applied, avoiding air bubbles and the vat was covered. The gel was run for 5 minutes at 60V and % mA/gel until the dyes separated, then switched to 120V and % mA/gel. Check for air bubbles rising inside the vat. After finishing the run (when the second blue dye was coming out of the gel), the voltage was lowered and the source switched off. The gel was carefully removed from the plate holder and placed in a glass vat with fixing solution for 10 minutes or up to 24 hours, shaking gently. When the fixative solution had completely covered the gel (approximately 150 ml), the fixative solution (save) was removed from the vat and 150 ml of colouring agent (silver nitrate) was added, avoiding spilling it on the gel. It was stirred gently for 10 minutes. The colouring solution was removed, washed quickly in MQ water and again with MQ water for 2 minutes, shaking gently, the MQ water was discarded, 150 ml of developer solution (sodium hydroxide) was added along with 1 ml of formaldehyde. It was stirred gently for about 10 minutes, observing the intensity of the bands, the developer solution was discarded and the fixative solution (which had been saved) was added when the gel was stained.

## 4.4 Analysing the data

Considering the essentially qualitative nature of the data and the innovative technological nature of the products generated, the statistical analyses carried out were limited to measures of central tendency and variance calculated in spreadsheet programmes (such as Microsoft Excel).

# CHAPTER 5

## RESULTS AND DISCUSSION

As presented in the methodology, genetic obesity markers can involve different approaches, which will analyse aspects of phenotypic markers (based on symptoms and comorbidities associated with genes involved in genetic obesity), hereditary markers (construction and analysis of heredograms) and molecular markers (computational prediction and experimental training). Before starting to discuss the results themselves, it should be emphasised that even though there is a genetic component, obesity is polygenic and multifactorial, so there is no expectation of identifying a single marker, but rather a set of aspects that direct nutritional conduct towards monitoring based on genetic individuality.

In the first stage, 119 questionnaires were analysed, answered by initially obese individuals from 14 Brazilian states and the Federal District, who had undergone bariatric surgery and had or had not regained weight after surgery. The assumption of this analysis is that, since all the individuals underwent multi-professional follow-up after surgery, the group with weight regain has a higher density of individuals with genetic alterations associated with obesity. The general profile of the research participants can be seen in Table 2 and Graph 1.

**Table 2.** Profile of interviewees according to gender, age and when they realised they were overweight.

| Sex | Age | | | Total |
|---|---|---|---|---|
| | **17-29 years** | **30 - 45 years** | **46 - 60 years** | |
| **Female** | 30 | 69 | 9 | 108 |
| **Male** | 3 | 7 | 1 | 11 |
| **Total** | 33 | 76 | 10 | 119 |
| **At what point did you realise you were overweight?** | | | | |
| | Childhood | Adolescence | Adulthood | |
| **No. of participants** | 62 | 23 | 34 | 109 |

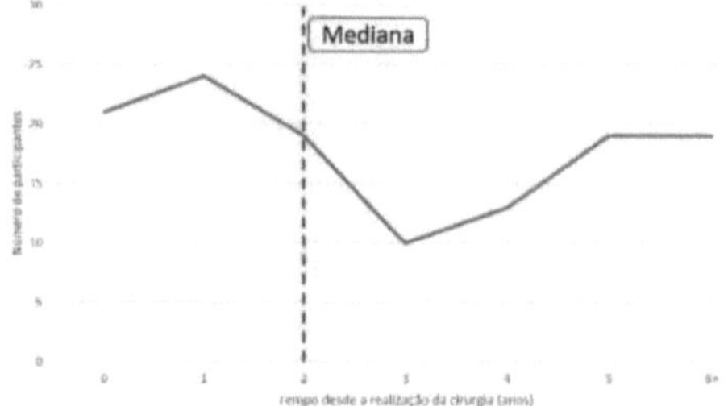

**Graph 1:** Frequency distribution of survey participants according to time since bariatric surgery.

As can be seen, the vast majority of participants were female (90.7%), while only 9.3% were male. With regard to age, the majority were between 30 and 45 years old (63.9%). When asked how long they had been living with the problem of excess weight, the majority (52.1%) said they had been aware of it since childhood. This is interesting because childhood and adolescent obesity is associated with both genetic and dietary factors, but we don't know about the association between differential genetic expression in adulthood and obesity. We also asked about the time that had elapsed since bariatric surgery, since this can affect the characterisation of weight gain. Graph 1 shows that the median time was 2 years, i.e. 50% of the participants had undergone bariatric surgery no more than 2 years before completing the questionnaire.

## 5. 1Phenotypical aspects      of genetic obesity.

At this stage, the aim was to identify some signs or symptoms (treated here as phenotypes) that could lead to the search for genetic polymorphisms

associated with obesity. Of the 119 questionnaires analysed, 8 were excluded due to errors in completing the form or lack of essential information. Of the remaining 111, 33 individuals did not report any of the phenotypes or did not want to answer and 78 participants claimed to experience some of the phenotypes associated with the genes involved in genetic obesity. The complaints reported by the participants, according to the groups with and without weight gain after the surgical procedure, are shown in Graph 2:

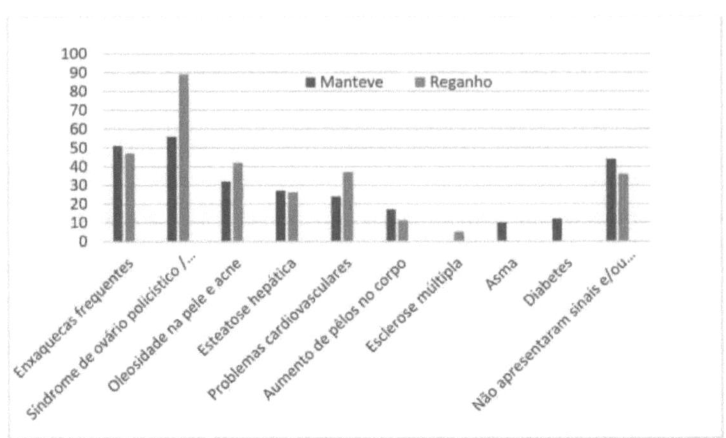


**Graph 2:** Phenotypes associated with genes involved in obesity, according to the percentage of complaining participants in the groups with and without post-surgical weight gain.

Considering only the 78 individuals who reported one of the phenotypes described, 59 belonged to the group without weight gain and 19 returned to approximately the same weight as before surgery. The most frequently reported phenotype in both groups was polycystic ovary syndrome which, being exclusive to women, reflects the high number of females in this study compared to males. Considering the sample size in each group, no significant differences were found between the phenotypes observed in the two groups. Despite this, almost 90% of the participants who gained weight reported polycystic ovary syndrome and none of them reported asthma or diabetes, which was observed exclusively in the group that did not gain weight.

According to Campos et al (2016), being overweight is linked to the presence of various chronic diseases, the most common of which are hypertension and diabetes. In this study, diseases often linked to obesity and eating habits, such as diabetes and cardiovascular problems, had a low number of responses in the group that did not gain weight. The low percentage of the diabetes phenotype in the group that did not gain weight reinforces this relationship between excess weight and diabetes; moreover, the absence of reports of the phenotype in the group that did gain weight suggests the occurrence of other non-dietary factors that determine weight, such as genetic factors. Thus, post-surgical weight gain without the occurrence of diabetes could lead to research into genetic polymorphisms associated with obesity, and a systematic analysis with a larger number of participants and a "gold standard" for molecular diagnosis would be important to validate this hypothesis.

## 5.2  Hereditary aspects of obesity and genetic obesity.

In addition to signs and symptoms suggestive of genetic obesity, the heredity profile of the obesity phenotype was investigated among the 119 research participants mentioned above. However, 25 of the 119 participants (21 per cent) had to be excluded due to errors in filling in the form or lack of essential information for the construction of heredograms. Two examples of the heredograms constructed in this study are shown in Figure 1:

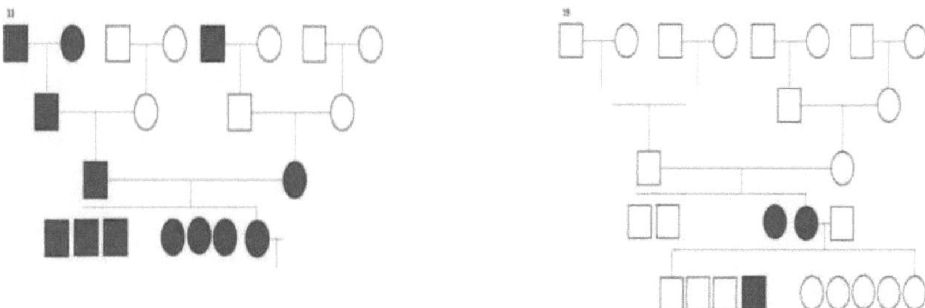

**Figure 1** - Examples of heredograms. On the left is a heredogram with a profile suggesting genetic inheritance for the "obesity" phenotype. On the right is a heredogram with no profile suggesting genetic inheritance.

No genetic inheritance patterns were observed in 69 of the remaining 94 participants (73.4%), suggesting a predominantly dietary aetiology for obesity. On the other hand, 25 participants (26.6%) had a profile suggestive of hereditary obesity. Table 1 summarises this data:

**Table 1.** Distribution of study participants according to heredogram analysis and weight gain after surgery.

| Analysing Heredogram | Maintained weight after surgery | | Total |
|---|---|---|---|
| | Yes | No | |
| Suggested heredity | 17(18,1%) | 8 (8,5%) | 25 (26,6%) |
| Not suggestive of | 54 (57,4%) | 15(16%) | 69 (73,4%) |

| heredity | | | |
|---|---|---|---|
| Total | 71 (75,5%) | 23 (24,5%) | 94(100%) |

Analysing the heredograms, it can be seen that more than half of the participants (57.4%) maintained their weight after the surgical procedure and had no family history suggestive of inherited obesity, a profile that is quite compatible with obesity of alimentary etiology, whose monitoring with a balanced diet and physical exercise is satisfactory. In 18.1% of cases, despite surgical success, it is believed that there is a greater possibility of weight regain due to the hereditary component, which is why it would be interesting to monitor the first group more closely and, if necessary, follow up with a diet based on genetic knowledge.

Almost 1/6 (16%) of the individuals showed weight gain after surgery, but no genetic inheritance pattern was identified. One possible cause for this relationship is "*de novo*" mutations, genetic polymorphisms that appear in an individual without being inherited. Various factors can cause new mutations in an individual, from food to solar radiation, and in these cases it is very difficult to identify where (in the genetic material) the mutation occurred without in-depth knowledge of cellular metabolism and appropriate genetic markers (which will be discussed in the next sections of this paper). Finally, a dominant or recessive inheritance pattern and weight gain were observed in 8.5% of cases. This profile is compatible with obesity of genetic etiology and requires a more in-depth clinical/laboratory investigation, as there are several known genetic syndromes involved in obesity that have specific and individualised diet plans that respect the genetic inheritance of each individual. It is important to note that genetic obesity does not imply heredity, since genetic obesity involves polymorphisms present in the genetic material, DNA and RNA, regardless of whether these polymorphisms are or have been passed down through the generations. In this study, all the participants were asked about genetic testing or guidance in this regard and none reported receiving any guidance (data not shown).

Of the 25 heredograms in which it was possible to observe a pattern of genetic inheritance, 8 had a pattern compatible with dominant inheritance (with the obesity phenotype observed in three or more consecutive generations) and 14 compatible with recessive inheritance (phenotype observed in two or more alternate generations). In 3 heredograms it was not possible to identify a pattern of genetic inheritance, as they showed both dominant and recessive characteristics, according to the information collected. Table

2 summarises these results:

**Table 2.** Distribution of heredograms according to the most compatible inheritance pattern identified (dominant or recessive).

| Heredogram suggestive of genetic inheritance | Did you maintain your weight after surgery? | | Total |
| --- | --- | --- | --- |
| | YES | NO | |
| **Recessive** | 10 | 4 | 14 |
| **Dominant** | 5 | 3 | 8 |
| **Total** | 15 | 7 | 22 |

Inheritance patterns with intermediate phenotypes, differences in penetrance and expressiveness or multifactorial aspects were not analysed due to the limitations of the data collected. No heredograms suggestive of extra-nuclear or sex-linked inheritance were observed. It was possible to observe that regardless of weight gain or not, most of the heredograms suggest recessive genetic inheritance. This is in line with data in the literature showing that most mutant alleles are recessive. In addition, this data helps to identify a marker for obesity, since it directs the search for polymorphic genes with relevant allele frequency and known phenotypic aspects.

Another important aspect of this study is that the assumption of genetic obesity is based solely on the occurrence of obese parents or siblings. It is relatively common for people (professionals specialising in the area or not) to attribute the cause of an individual's obesity to genetic inheritance, simply because they have parents and/or siblings who are also obese. This view is mistaken, as they probably live in the same house and may have similar eating patterns, which would characterise dietary obesity and not necessarily genetic. From this perspective, we present the construction of heredograms that include at least three generations as a simple, practical and functional tool that can help identify patients with genetic obesity and direct personalised treatment strategies.

Finally, by knowing the inheritance pattern of some polymorphic genes for genetic obesity and the information obtained from the heredograms, it is possible to identify individuals with a greater need for genetic research, as well as focus the study on the most likely genes, significantly reducing costs.

## 1.3. Data mining of polymorphisms associated with genetic obesity.

As discussed above, in this study the identification of obesity markers with a genetic etiology involved the use of three distinct and complementary approaches, of which the identification of genetic polymorphisms and the development of molecular analysis tools represented the last stage. Initially, the *data mining* strategy used was a bibliographic search for polymorphisms associated with obesity, followed by their identification in biological sequence databases and a new bibliographic search based on the referenced data. Biological data was extracted mainly from the dbSNP database, which contains information on genetic polymorphisms. This database is a tool that has been widely used in a number of very recent and innovative studies, such as the study by Smith and Mclean (2017) which investigated alterations in any of the five keratin genes that, through dominant negative heterozygous mutations, can lead to a rare dermal disease called congenital pachyonychia. This same tool was used in a study by Yuan, et al (2016) in which MTHFR gene variants were evaluated as a risk factor for Parkinson's disease in the Chinese population. The study by Suresh, Venkatesh and Tsutsumi (2016) emphasised the importance of studying SNPs to understand the genetic basis of hepatocellular carcinoma. Figure 2 shows the dbSNP homepage and illustrates how the information is initially presented.

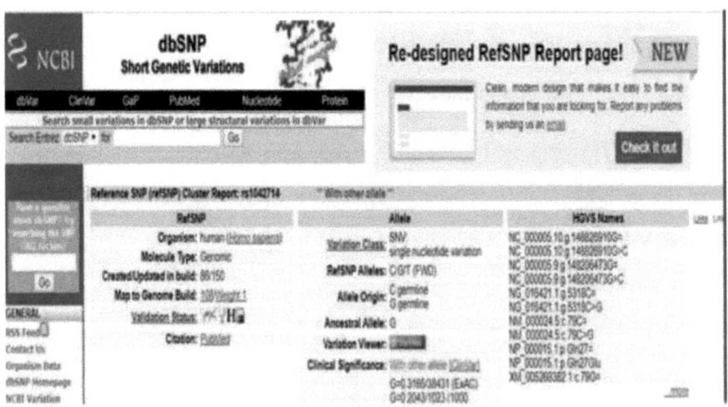

**Figura 2.** dbSNP homepage (accessed on 05/09/2017).

In the database, all the sequences are located using a polymorphic identifier code (rs[0-9]n) linked to a wide range of information such as allele and genotype frequencies, epidemiology, mutation site, bibliographic references and complete sequences for each of these genes. From this, the most important genes were selected and from these, the polymorphic sites with the greatest potential for developing molecular tools. A total of 50 polymorphisms were selected from 36 genes cited in the literature as being associated with obesity. Table 4 shows the genes selected by *data mining*, their respective polymorphic identifiers and mutations associated with obesity.

**Chart 3:** Genes and genetic polymorphisms associated with obesity studied in this study.

| GENE | POLYMORPHIC IDENTIFIER | CHANGE | GENE | POLYMORPHIC IDENTIFIER | CHANGE |
|---|---|---|---|---|---|
| ADIPOR2 | rs55789623 | 395 G/C | KCNJ11 | rs5210 | 501 G/A |
| ADRB2 | Rs1042714 | 329 G/C | KCTD15 | Rs11084753 | A/G |
| APM1 | rs2241766 | 501 T/G | LEP | rs2167270 | 501 A/G |
| APOA2 | rs5082 | 501 T/C | | Rs7799039 | A/C/G |
| BCDIN3 | Rs7138803 | 501 A/G | LIPC | rs2070895 | 501 G/A |
| CD36 | rs1761667 | 501 A/G | MC4R | Rs17782313 | 501 C/T |
| CHST8 | Rs29941 | 501 T/C | MGAT1 | rs4285184 | 501 A/G |
| CYP1A2 | Rs762551 | 501C/A | MTHFR | Rs1801133 | 501 C/T |
| DRD2 | rs1800497 | 501 C/T | NEGR1 | RS2815752 | 501 C/T |
| FABP2 | rs1799883 | 501 A/G | | Rs2568958 | 501 C/T |
| FKBP5 | rs1360780 | 501 T/C | NPY | rs164147 | 501 A/C |
| FTO | rs9930506 | A/G | PARD | rs2016520 | 274 A/G |
| | Rs9939609 | A/T | PPARgamma | Rs1801282 | 501 C/G |
| | rs1421085 | C/T | PRL | RS4712652 | 501 A/G |
| GHR | rs4410646 | 501 C/A | SH2B1 | RS7498665 | 501 G/A |
| | rs4898743 | 501 C/A/T | TCF7L2 | rs4132670 | 301 C/T |
| GIPR | rs2287019 | C/T | | rs11196175 | 256 C/T |
| HNF4A | rs1885088 | 301 A/G | | rs12255372 | 294 A/G/T |
| IL6 | Rs1800795 | C/G | | rs7903146 | 346 C/T |
| INSIG2 | RS7566605 | 501 C/G | | rs11196236 | 256 C/T |
| IRS1 | rs2943641 | 501 C/T | | rs6548238 | C/T |
| IRX3 | rs8053360 | 311 C/T | TMEMB18 | RS7561317 | 501 A/G |
| | rs3751723 | 201 A/C | | rs4854344 | 501 G/A |
| | rs12445085 | 501 G/T | TNFalpha | Rs1800629 | 501 A/G |
| | rs1126960 | 301 A/C/G/T | TRIB2 | rs1057001 | 501 T/A |

As well as polymorphic sites, information was extracted on allele frequencies in the different populations studied. Since the number of polymorphic alleles is large, it is necessary to select those with the greatest potential for population surveys, and in this sense allele frequency is an important parameter. Figure 3 illustrates this analysis:

**Figura 3.** Analysis of allele frequencies available on dbSNP.

From the *data mining*, the polymorphic sequences were extracted in FASTA format following the format of the database itself (polymorphic nucleotide represented by the degenerate code and 500 nucleotides 5' and 3', totalling a fragment of 1001 nucleotides). Figure 4 illustrates the FASTA file.

**Figura 4.** Example sequence extracted from dbSNP.

The sequences identified in the previous stage underwent local alignment and global alignment. Local alignment made it possible to identify duplicate genes or similar genetic sequences, which would make accurate molecular diagnosis impossible. To do this, the BLAST programme (Figure 5) was used to compare the polymorphic sequence with the "Nucleot/de *colection (nr/nt)*" database for the "Human *(taxid:9606)"* species, both available on the NCBI. This comparison enabled the sequences corresponding to the polymorphisms studied to be identified within the database.

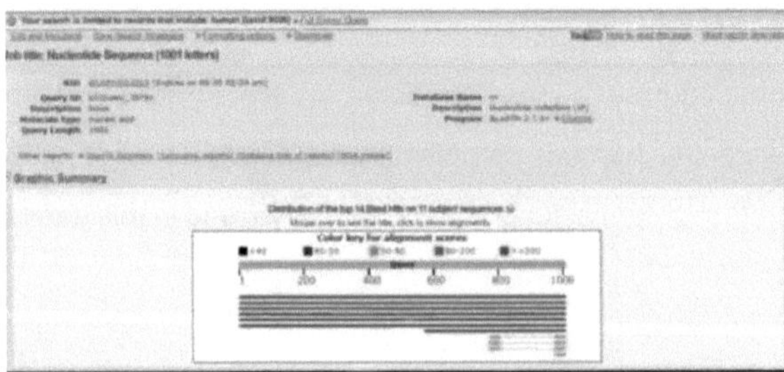

**Figure 5 -** Example of the result of an online BLAST

Among these sequences, we found homologous sequences related to polymorphisms which, through global alignment using the *Clustal 2.1* programme, identified the conserved and variable regions of the genes. Of the 50 polymorphisms selected, 11 were excluded in the BLAST and/or CLUSTAL 2.1 stages. Of these 11 genes, 6 presented problems in BLAST, such as a lack of sequences that aligned with the polymorphic gene and alignments with few nucleotides, which were considered to be poor quality and unreliable. The other 5 genes were excluded because, when submitted to the CLUSTAL 2.1 programme, they did not differentiate well between conserved and variable regions and/or had too many insertions in the polymorphic region. Table 4 below shows the 39 polymorphisms of the remaining 31 genes, together with the number of homologous sequences that aligned with the gene through local alignment by BLAST.

**Table 4.** Number of sequences identified in the human genome from local alignment of obesity-associated polymorphisms.

| GENE | POLYMORPHIC IDENTIFIER | N° SEQUENCES (BLAST) | GENE | POLYMORPHIC IDENTIFIER | N° SEQUENCES (BLAST) |
|---|---|---|---|---|---|
| ADIPOR2 | rs55789623 | 14 | MC4R | Rs17782313 | 2 |
| ADRB2 | RS1042714 | 31 | MGAT1 | rs4285184 | 104 |
| APM1 | rs2241766 | 29 | MTHFR | Rs1801133 | 28 |
| APOA2 | rs5082 | 14 | | RS2815752 | 11 |
| BCDIN3 | Rs7138803 | 8 | NEGR1 | RS2568958 | 102 |
| CD36 | rs1761667 | 3 | NPY | rs164147 | 9 |
| CHST8 | Rs29941 | 107 | PARD | rs2016520 | 28 |
| CYP1A2 | Rs762551 | 24 | PPARgamma | Rs1801282 | 21 |
| DRD2 | rs1800497 | 23 | PRL | RS4712652 | 1 |
| FABP2 | rs1799883 | 14 | SH2B1 | RS7498665 | 64 |
| FKBP5 | rs1360780 | 100 | | rs4132670 | 2 |
| | rs4410646 | 2 | | rs11196175 | 2 |
| GHR_1 | rs4898743 | 103 | | rs12255372 | 2 |
| IRS1 | rs2943641 | 56 | TCF7L2 | rs7903146 | 3 |
| | rs12445085 | 5 | | rs11196236 | 2 |
| IRX3 | rs1126960 | 17 | | RS7561317 | 2 |
| KCNJ11 | rs5210 | 23 | TMEMB18 | rs4854344 | 102 |
| KCTD15 | Rs11084753 | 3588 | TNFalpha | Rs1800629 | 59 |
| LEP1 | rs2167270 | 9 | TRIB2 | rs1057001 | 9 |
| LIPC | rs2070895 | 43 | | | |

**5.4 Development of a molecular diagnostic tool for genetic obesity in a virtual machine.**

The 39 polymorphisms selected in the previous stage were used to build restriction maps, with the aim of identifying restriction enzymes capable of selectively cutting polymorphic fragments associated with obesity. The Rebase database previously described in the methodology was used for this construction. The main parameters for constructing the maps were: selecting only enzymes with sites that cut five times or less, displaying a frequency summary table with a list of enzymes that do not cut the fragment being analysed, including enzymes with degenerate recognition with large recognition sites and all enzyme schizomers. Figures 6 and 7 represent restriction maps according to the parameters and database mentioned:

**Figure 6:** Restriction map of a fragment of the TCF7L2 gene (rs7903146 polymorphism). The 5' flanking region of the polymorphism being analysed is highlighted in blue.

**Figure 7:** Restriction map of a fragment of the APOA2 gene (rs5082 polymorphism). In the figure, the 5' flanking region of the polymorphism being analysed is highlighted in blue.

When analysing all the restriction maps, it was observed that of the 39 polymorphisms, 33 have different enzymes when comparing the two maps. The presence of different enzymes between the two maps is what enables us to differentiate, during experimental diagnosis, which patient has this polymorphism and which does not, which is why these sequences were kept in the study. The average number of enzymes per polymorphism was approximately 6 and overall it was noted that the minimum number of enzymes identified per polymorphism was 1 and the maximum number was 26 enzymes. Due to the large number of enzymes identified by polymorphism, it is believed that it is easier to assemble markers due to the greater possibility of choice, which can be conditioned by economic flexibility, feasibility of application and availability of the enzyme on the market.Once the restriction enzymes had been selected, the next step was to analyse the conserved region of the genomic fragments, more specifically the conserved 5' and 3' ends of the 1OO1pb fragment (sequence originally extracted in the *data mining* phase). These regions were analysed to design PCR primers using the GeneTool program (Lyion, 2000) and after simulations and in *silico* tests, the following criteria were adopted: a) the primers must have an average of 20nt with a *melting* temperature of around 55 - 60°C; b) both primers (sense and antisense) must have equal or very close *melting* temperatures, they must not dimerise, assume a ring conformation or present instability problems; c) the amplified product must not be less than 200nt, it must have the polymorphic site associated with obesity and this must not occupy the central position of the fragment. Figure 8 shows this analysis:

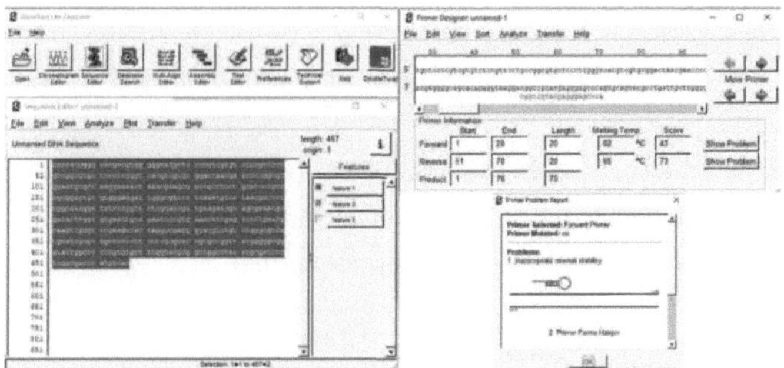

**Figure 8:** Design of PCR primers using the GeneTool tool.

Using these analysis criteria, it was possible to design sense and antisense primers for all the polymorphic sequences under study. This result was somewhat unexpected and given the large number of potential obesity markers remaining, a new analysis was carried out by local alignment of the primers, using the BI_AST tool, already described, and the human genome, in order to identify possible non-specific links. Finally, virtual gels were constructed simulating the conditions of amplification, enzymatic digestion and electrophoresis of the polymorphisms studied, using the GeneTool tool. Figure 9 illustrates some of these gels:

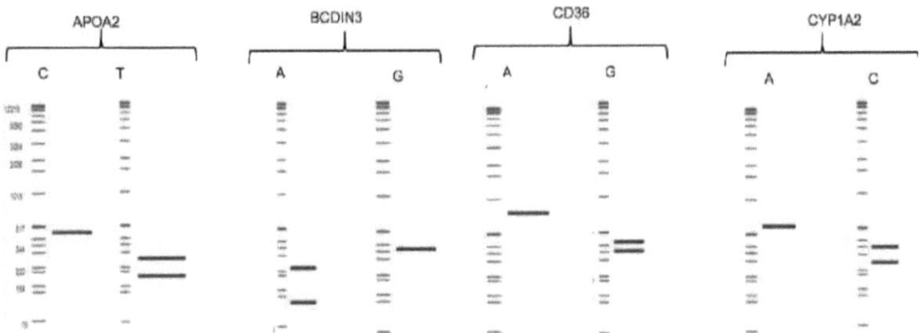


**Figure 9:** Virtual gels simulating the experimental conditions for the molecular diagnosis of obesity of genetic etiology. The figure shows simulated polyacrylamide gels of DNA fragments amplified by PCR using the primers constructed in this project and digested with specific restriction enzymes identified according to the methodology presented above. The PCR, enzyme digestion and electrophoresis conditions are simulation parameters and represent the experimental situation with high fidelity.

Direct analysis of the virtual gels clearly shows that using PCR-RFLP techniques it was possible to characterise and identify patients with genetic obesity polymorphisms in 33 of the 50 polymorphisms initially studied (66% of the polymorphisms).

## 5.5 Experimental training and economic viability.

Specific training in bioinformatics and molecular biology was carried out to ensure the safety and quality of the work. The training was carried out at the Federal University of Piauí in the Biology Department at the Bioinformatics and Genomics Laboratory, under the supervision of Professor Daniel Barbosa Liarte. In the first approach, the routine of protocols and programmes currently used by our group was analysed, new databases were inserted and adapted to the research, and programmes for format conversion and data extraction were optimised in order to speed up the analysis procedure. In the area of molecular biology, human DNA was extracted and polyacrylamide gels were made and used for electrophoresis. In both approaches it was possible to absorb a huge amount of knowledge, so that the assimilation between theory and practice was well established.

**Figure 10.** Demonstration of equipment used during training

In addition, a brief analysis was made of the economic potential of using these techniques and tools developed in this work, in comparison with the proposals for molecular diagnosis of obesity available on the market. To this end, a spreadsheet was drawn up of costs relating to DNA collection and extraction procedures, PCR-RFLP and polyacrylamide gel electrophoresis. These costs were compared with the cost of DNA sequencing, a technique traditionally used to diagnose genetic obesity. Although the costs vary significantly between suppliers and genes investigated, the protocol developed is 5 - 10X cheaper than the current commercial alternative, which is highly significant, given that there are services of this standard budgeted at R$13,000.00 in private companies. This high financial cost leads many patients not to undergo this assessment, and they are often subjected to surgery, diets and aesthetic procedures that have no positive impact on weight loss, since the conditioning factor of this obesity may be linked to genetic material.

# CHAPTER 6

## CONCLUSION

Through this research, it was possible to identify phenotypic and molecular markers for obesity, which, when used in combination, direct the professional's conduct towards a low-cost and effective diagnosis of genetic obesity, allowing for the creation of more specific and individualised diet plans. In this way, we can see the enormous viability of using these markers in the multiprofessional monitoring of obese patients.

## REFERENCES


ABESO. Brazilian Association for the Study of Obesity and Metabolic Syndrome. **Brazilian obesity guidelines 2009/2010.** 3. ed. Itapevi, São Paulo, 2009.

ABESO. Brazilian Association for the Study of Obesity and Syndrome Metabolic. **Obesity map.** Available at: http://www.abeso.orq.br/atitudesaudavel/mapa-obesidade. Accessed on: 02 May 2017. 2010

ALTSCHUL, S. F. et al. Gapped BI_AST and PSI-BLAST: a new generation of protein database search programmes. **Nucleic acids research, v.** 25, n. 17, p. 3389-3402, 1997.

BRAZIL. Ministry of Health; Pan American Health Organisation. **Health promoters: experiences from Brazil.** Brasília- DF. Ministry of Health Publishing House, 2006a.

BRAZIL. Ministry of Health. General Coordination of Food and Nutrition Policy. Primary Care Department, Health Care Secretariat. **Food guide for the Brazilian population: promoting healthy eating.** 2006b.

BRAZIL. Ministry of Health. **Vigitel Brazil 2014: surveillance of risk and protective factors for chronic diseases by telephone survey.** Health Surveillance Secretariat, Department of Surveillance of Non-Communicable Diseases and Health Promotion. - Brasília: Ministry of Health, 2015.

BRAZIL. Ministry of Health. **Vigitel Brazil 2015. Supplementary Health: surveillance of risk and protective factors for chronic diseases by telephone survey.** National Supplementary Health Agency. - Brasília: Ministry of Health, 2017a.

BRAZIL. United Nations Organisation in Brazil. **Overweight and obesity on the rise in Brazil, says FAO and PAHO report.** Available at: https://nacoesunidas.org/aumentam-sobrepeso-e-obesidade-no-brasil- apontarelatorio-de-fao-e-opas/. Accessed on: 02 May 2017. 2017b.

BRODOWSKI, J. et al. Searching for the relationship between the parameters of metabolic syndrome and the rs17782313 (T> C) polymorphism of the MC4R gene in postmenopausal women. **Clinicai Interventions in Aging,** v. 12, p. 549, 2017.

CANTALICE, A. S. C et al. Persistence of metabolic syndrome in overweight children and adolescents according to two diagnostic criteria: A longitudinal study. **Rev. de Medicina (Ribeirão Preto. Online),** v. 48, n. 4, p. 342348, 2015.

CAMPOS, J. R. et al. The impact of fluctuating weight on cardiovascular risk factors in obese women. **HU Revista,** v. 41, n. 3 and 4, 2016.

COUTINHO, W.; DUALIB, P. Etiology of obesity. **Revista da ABESO,** v. 30, n. 30, 2007. DAMIANI, D.; DAMIANI, D.; OLIVEIRA, R.G. Obesity-genetic or environmental factors. **Modern Paediatrics,** v. 38, n. 3, p. 57-80, 2002.

DE MORAIS, T. M. D. M. et al. Booklet for adults with metabolic syndrome: Proposal for educational technology for health promotion. **Proceedings of the Seminar on Technologies Applied to Education and Health,** 2017.

FAYYAD, U.; SHAPIRO, G.; SMYTH, P. The KDD process for extracting useful knowledge from volumes of data. **Communications of the ACM,** v. 39, n. 11, p. 2734, 1996.

FERNANDES, A. E.; FUJIWARA, C. T. H.; MELO, M. E. Genetics: A Common Cause of Obesity. Obesity and Metabolic Syndrome Group of the Hospital das Clínicas of the Faculty of Medicine of the University of São Paulo. ABESO, 2011.

FRIGERI, H. R. **Genetic variability and sequencing of genes associated with type 2 diabetes mellitus and obesity.** 2015. p. 42, 52. Thesis (Doctorate in Pharmaceutical Sciences) - Federal University of Paraná, Curitiba, 2015.

FUJII, T. M. M.; MEDEIROS, R.D. and YAMADA, R. Nutrigenomics and nutrigenetics: important concepts for nutrition science. **Nutrire Rev. Soc. Bras. Aliment. Nutr,** v. 35, n. 1, 2010.

GENOVIVE. **Genetic profile report for weight control based on diet and exercise.** Available at: http://qenovivebrasil.com.br/dieta.php. Accessed on: 02 May 2017. 2011.

GLOY, V. L. et al. Bariatric surgery versus non-surgical treatment for obesity: a systematic review and meta-analysis of randomised controlled trials. **Bmj,** v. 347, p. f5934, 2013.

HALPERN, Z. S.C; RODRIGUES, M. D. B.; COSTA, R.F. Physiological determinants of weight control and appetite. **Rev Psiq Clin,** v. 31, n. 4, p. 150-3, 2004.

BRAZILIAN INSTITUTE OF GEOGRAPHY AND STATISTICS. **National Health Survey 2013: access to and use of health services, accidents and violence.** Rio de Janeiro: IBGE; 2015.

LANDEIRO, F. M.; DE CASTRO QUARANTINI, L. Obesity: neural and hormonal control of eating behaviour. **Rev. de Ciências Médicas e Biológicas,** v. 10, n. 3, p. 236-245, 2011.

LAYON, M. "GeneTool 1.0: Update 4." Biotech Software & Internet Report: **The Computer Software Journal for Scient, v.** 1, n.6, p. 261-264. 2000.

LOPES, J. F. et al. Effect of gradual changes in physical exercise and diet on the body composition of obese people. **Arquivos de Ciências da Saúde,** v. 24, n. 1, p. 9397, 2017.

MACHADO, I. C; KIRSTEN, V. R. Adherence to nutritional treatment of adult patients treated at a clinic in Santa Maria-RS. **Disciplinarum Scientia| Saúde,** v. 12, n. 1, p. 81-91, 2016.

MARTINS, I. S.; MARINHO, S. P. The diagnostic potential of centralised obesity indicators. **Rev. de Saúde Pública, v.** 37, n. 6, p. 760-767, 2003.

MONTEIRO, R. C. A.; RIETHER, P. T. A.; BURINI, R. C. Effect of a mixed programme of nutritional intervention and physical exercise on the body composition and eating habits of obese climacteric women. **Rev. de Nutrição,** p. 479489, 2004.

NOVAIS, P. F. S. et al. Evolution and classification of body weight in relation to the results of bariatric surgery: Roux-en-Y gastric bypass. **Arquivos Brasileiros de Endocrinologia & Metabologia,** p. 303-310, 2010.

OCHIONI, A. C. **Analyses of polymorphisms of inflammatory mediator genes involved in obesity.** 2016.p. 35-90. Dissertation (Master's in Cellular and Molecular Biology) - Osvaldo Cruz Institute, Rio de Janeiro, 2016.

ORDOVAS, J. M.; CORELLA, D. Nutritional genomics. **Annu. Rev. Genomics Hum. Genet.,** v. 5, p. 71-118, 2004.

PLOMIN, R. et al. **Genetics of Behaviour-5ª Edition.** Artmed Editora, 2016.

POPKIN, B. M. etal. The nutrition transition in China: a cross-sectional analysis. **European journal of clinical nutrition, v.** 47, n. 5, p. 333-346, 1993.

PROSDOCIMI, F. et al. Bioinformatics: user's manual. **Biotecnologia Ciência & Desenvolvimento,** v. 29, p. 12-25, 2002.

RICARTE, K. M. P. et al. Relationship between nutritional status and metabolic syndrome in adolescents from the semi-arid region of Piauí. **Ciência, Cuidado e Saúde,** v. 16, n. 2, 2017.

RIBEIRO, S. M. L. et al. Leptin: aspects on energy balance, physical exercise and effort amenorrhoea. **Arquivos Brasileiros de Endocrinologia e Metabologia,** v. 51, n. 1, p. 11, 2007.

ROBERTS, R.J. et al. REBASE-a database for DNA restriction and modification: enzymes, genes and genomes. **Nucleic Acids Res.** V.43, p. 298-299. 2015.

ROCHA, S. A. M. T; VIANA, E. S.M; MOREIRA, BORONI, A. P. Impacto da intervenção nutricional em indivíduos com excesso peso atendidos na clínica escola de uma instituição de ensino superior. **ANAIS SIMPAC,** v. 6, n. 1, 2016.

ROJANO-RODRIGUEZ, M. E. et al. Leptin receptor gene polymorphisms and morbid obesity in Mexican patients. **Hereditas,** v. 153, n. 1, p. 2, 2016.

ROMAGNA, E. S., SILVA, M. C. A. D. E BALLARDIN, P. A. Z. Prevalence of overweight and obesity in children and adolescents at a basic health unit in Canoas, Rio Grande do Sul, and comparison of nutritional diagnosis between the CDC 2000 and WHO 2006 charts. **Sei Med,** v. 20, n. 3, p. 228-31, 2010.

SCHUCH, J. B.; VOIGT, F.; MALUF, S. W. E ANDRADE, F. M. Nutrigenetics: the interaction between eating habits and individual genetic profile. **Rev. Brasileira de Biociências,** v. 8, n. 1, 2010.

SMITH, FRANCES JD; MCLEAN, W. H. Keratin 6b variant p. Gly499Ser reported in delayed-onset pachyonychia congenita is a non-pathogenic polymorphism. **The Journal of Dermatology,** 2017.

BRAZILIAN SOCIETY OF ENDOCRINOLOGY AND METABOLISM. **The metabolic syndrome.** Available at: <https://www.endocrino.org.br/a-sindrome-metabolica/>. Accessed on: 09 May 2017. 2016.

SOUZA, E. B. Nutritional transition in Brazil: analysis of the main factors. **Cadernos UniFOA,** v. 5, n. 13, p. 49-53, 2017.

SURESH, P. S.; VENKATESH, T.; TSUTSUMI, R. Mining of single nucleotide polymorphisms in the 3'untranslated region of liver cancer-implicated miR-122 target genes. **Annals of translational medicine, v.** 4, n. 5, 2016.

THOMPSON, J. D. et al. The CLUSTAL_X Windows interface: flexible strategies for multiple sequence alignment aided by quality analysis tools. **Nucleic acids research,** v. 25, n. 24, p. 4876-4882, 1997.

VERMA, P, A, et al. The rs2070895 (-250G/A) Single Nucleotide Polymorphism in Hepatic Lipase (HL) Gene and the Risk of Coronary Artery Disease in North Indian Population: A Case-Control Study. **Journal of clinical and diagnostic research: JCDR,** v. 10, n. 8, p. GC01, 2016.

WANG, J. et al. Association of rs12255372 in theTCF7L2 gene with type 2 diabetes mellitus: a meta-analysis. **Brazilian Journal of Medicine and Biological Research,** v. 46, n. 4, p. 382-393, 2013.

WHO. World Health Organisation. Obesity preventing and managing the global epidemic. Report of a WHO consutation. **World Health Organ. Tech.Rep. Ser.,** V. 894, p 1-7, 1-253, 2000.

YUAN, L. et al. Association of the MTHFR rs1801131 and rs1801133 variants in sporadic Parkinson's disease patients. **Neuroscience letters,** v. 616, p. 26-31, 2016.

**APPENDIX**

**Appendix A**

MINISTRY OF EDUCATION FEDERAL
UNIVERSITY OF PIAUÍ
CENTRE FOR NATURAL SCIENCES - BIOLOGY DEPARTMENT

## Research project: Identification of phenotypic and molecular markers

# of genetic obesity.

PART I: CONSENT FORM.

This questionnaire is part of the research project mentioned above, under the responsibility of Nutrition student Juliana de Carvalho Passos and Genetics lecturer Daniel Barbosa Liarte. The study aims to **identify phenotypic and molecular characteristics pertinent to genetic obesity and thus develop diagnostic tools and control strategies based on the genetic profile.** The data will be collected completely anonymously, used only for research purposes and analysed in general rather than individually. If you feel uncomfortable with the questions in the questionnaire, you may withdraw from answering the questionnaire without any implications. Thank you in advance for your co-operation.

( ) I confirm my voluntary participation and agree to the terms presented above.

PART II: GENERAL DATA.

First name or surname: Sex: ( ) Male ( ) Female City of residence: State:

     Profession: ______________

Age (years): Marital status: ( ) Single ( ) Married or in a stable union

Do you have children? ( ) No ( ) Yes, how many?

BMI calculation:     Height: __ Weight: BMI:

Email or telephone contact (optional):

PART III: GENERAL ASPECTS ASSOCIATED WITH OBESITY.

How long have you been overweight?

    ( ) Since childhood ( ) Since adolescence ( ) Only in adulthood Do you practise physical activity?

        ( ) Yes, high or moderate intensity ( ) Yes, low intensity ( )I don't

            practise.

How often do you exercise?

        ( ) 3X a week or more ( ) Less than 3X a week

Do you think it's easy to put on weight? ( ) YES ( ) NO

Have you ever been on a diet prescribed by a nutritionist? ( ) YES ( ) NO

Have you ever undergone bariatric surgery? ( ) Yes ( ) No

How long ago did the surgery take place?

How do you rate your weight loss after surgery?

( ) The goal of the surgery was achieved ( ) I gained weight again after the surgery

Have you had any tests to assess your genetic profile associated with obesity? ( ) Yes ( )No

If so, which one?__________________

If yes, before having bariatric surgery? ( ) YES ( )NO

PART IV: IDENTIFICATION OF PHENOTYPES ASSOCIATED WITH OBESITY.

When you eat sugar, do you feel a quick surge of energy in your body?

( ) Yes ( )No ( ) Can't answer

Do you notice significant weight loss when you diet and exercise?

( ) Yes ( )No ( ) I can't answer How do you define your dietary pattern today?

( ) High in fat ( ) High in carbohydrates ( ) High in fibre and vegetables (

) High in protein ( ) Balanced.

What is your average blood glucose level?

Do you notice that even if you diet and exercise routinely, your LDL, HDL, triglyceride and total cholesterol levels don't drop?

( ) Yes ( )No ( ) Can't answer

Do you have any chronic diseases (e.g. diabetes, hypertension, etc.) or neural diseases (e.g. migraines, multiple sclerosis, arteriosclerosis, etc.)?

( ) Yes ( )No ( ) Can't answer

Do you often not feel full even after eating the recommended amount of food?

( ) Yes ( )No ( ) Can't answer

Do you have poorly distributed body fat, mainly concentrated on your stomach?

( ) Yes ( )No ( ) Can't answer

Do you have type 2 diabetes? ( ) Yes ( )No ( ) Can't answer

Do you have any of these conditions: insulin resistance, increased body hair, irregular menstruation, oily skin and acne, high blood pressure, cardiovascular problems, high cholesterol, polycystic ovary syndrome, hepatic steatosis (fatty liver)?

( ) Yes ( )No ( ) Can't answer

If yes, which one(s): _________________________________

Do you have any cases of cardiovascular problems in your family?

( ) Yes ( )No ( ) Can't answer

PART V: ASSEMBLING THE HEREDOGRAM

Please fill in the table below with the total number of biological relatives distributed according to weight (due to the nature of the survey, adoptive relatives should not be considered; if you don't know any of the fields, please leave them blank)

| Degree of kinship | Number of people | | | |
| --- | --- | --- | --- | --- |
| | Underweight | Ideal weight | Overweight | Obese |
| Children | | | | |
| Daughters | | | | |
| Spouse / partner | | | | |

| | | | | |
|---|---|---|---|---|
| Brothers | | | | |
| Sisters | | | | |
| Dad | | | | |
| Mum | | | | |
| Paternal grandfather | | | | |
| Maternal grandfather | | | | |
| Paternal grandmother | | | | |
| Maternal grandmother | | | | |
| Great-grandfather 1 (paternal grandfather) | | | | |
| Great-grandfather 2 (maternal grandfather) | | | | |
| Great-grandfather 3 (Paternal grandmother) | | | | |
| Great-grandfather 4 (maternal grandmother) | | | | |
| Great-grandmother 1 (paternal grandfather) | | | | |
| Great-grandmother 2 (maternal grandfather) | | | | |
| Great-grandmother 3 (paternal grandmother) | | | | |
| Great-grandmother 4 (maternal grandmother) | | | | |

Do you have relatives by adoption or breeding? If you do, please state the sex, degree of kinship and their classification (underweight, ideal weight, overweight or obese):

Auxiliary parameters for weight classification:

| | Abdominal circumference | BMI | Silhouette (see picture opposite) |
|---|---|---|---|
| Underweight | | <18,5 | Figure 1 |
| Ideal weight | | 18,5-24,9 | From 2 to 5 |
| Overweight | From 94 cm (men) and 80 cm (women) | 25 a 29,9 | From 5 to 7 |
| Obese | From 102 cm (men) and 88 cm (women) | From 30 | From 8 |

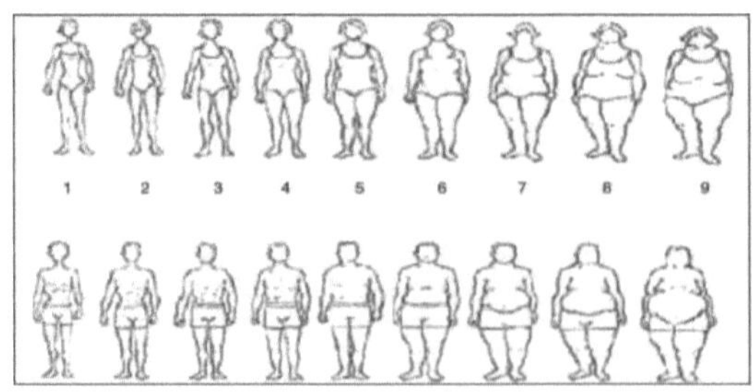

Printed by Books on Demand GmbH, Norderstedt / Germany